USES OF AI

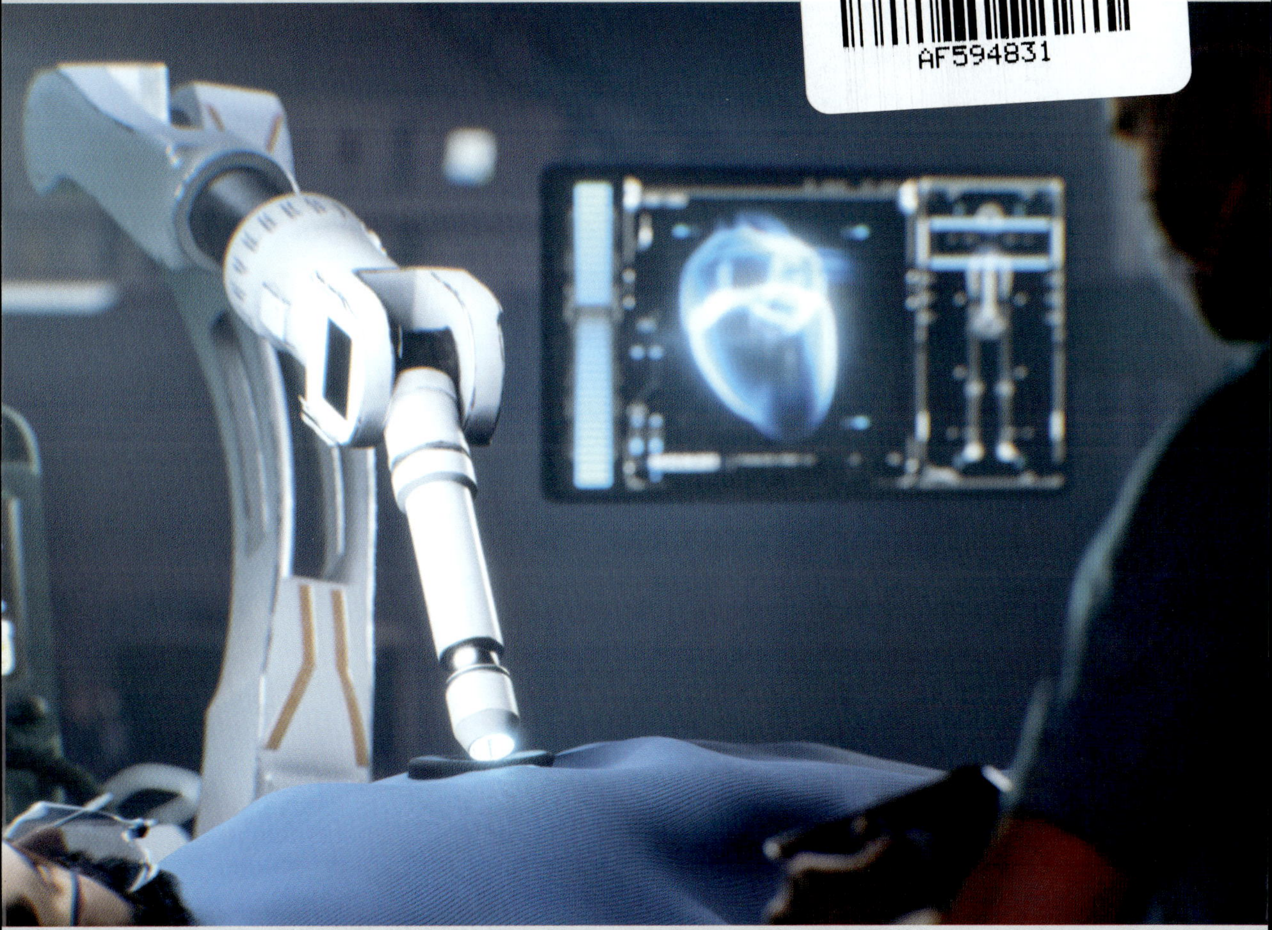

by Josh Gregory

Cherry Lake Press
Ann Arbor, Michigan

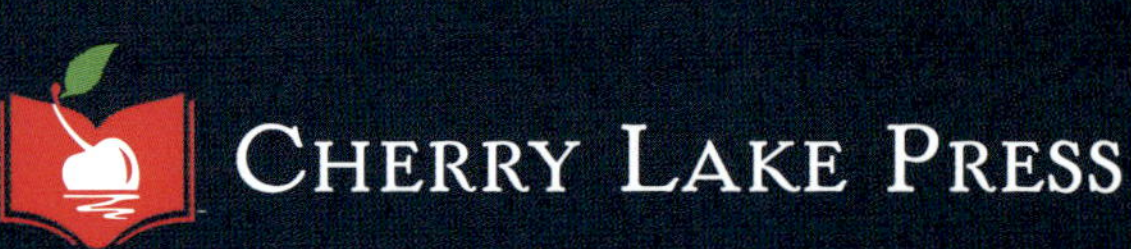

Published in the United States of America by Cherry Lake Publishing
Ann Arbor, Michigan
www.cherrylakepublishing.com

Reading Adviser: Beth Walker Gambro, MS, Ed., Reading Consultant, Yorkville, IL

Photo Credits: © sefa ozel/iStock, cover, title page; © Bekir Ugur/Dreamstime.com, 5; © Alejandro27/Dreamstime.com, 6; © Jjfarq/Dreamstime.com, 9; © Wrightstudio/Dreamstime.com, 11; © Narit Bualuang/Dreamstime.com, 13; © Dragoscondrea/Dreamstime.com, 14; © Altitudevs/Dreamstime.com, 17; © Yugong Luo/Shutterstock, 19; © Dashark/ Dreamstime.com, 21; © Neonic Flower/Shutterstock, 22; © Kinomaster/Dreamstime.com, 25; © Yuri Arcurs/Dreamstime.com, 27

Cherry Lake Press is an imprint of Cherry Lake Publishing Group.

Library of Congress Cataloging-in-Publication Data has been filed and is available at catalog.loc.gov.

Cherry Lake Press would like to acknowledge the work of the Partnership for 21st Century Learning, a Network of Battelle for Kids. Please visit Battelle for Kids online for more information.

Printed in the United States of America

ABOUT THE AUTHOR

Josh Gregory is the author of more than 200 books for kids. He has written about everything from animals to technology to history. A graduate of the University of Missouri–Columbia, he currently lives in Chicago, Illinois.

CONTENTS

Chapter 1

AI ALL AROUND US

Artificial intelligence, or AI, was once seen only in science-fiction stories. Computers in these stories could learn new things. They could think for themselves like humans do. It seemed like something that could only happen in the far future, if ever. Computers still aren't as intelligent or creative as humans. But AI technology is already all around us. It helps power countless devices and apps. Some are basic things we use every day. Others are advanced systems used in fields such as health care and finance. AI can help analyze **data**. It can generate content. It can **automate** tasks that once required humans to complete.

Have you ever wondered how Netflix or YouTube decide which videos to recommend to you? AI programs look at things such as videos you've watched before. They also look at which videos people with similar tastes watch. They use this information to guess which videos

AI algorithms are at work on most websites. They can enhance user experience.

you might want to watch next. Similar AI-powered systems help decide which songs play next in music apps. They decide which posts you will see in your social media feed. Amazon and other online stores also rely on these kinds of systems. They use them to recommend products to their customers.

Other AI systems figure out which ads you'll see when enjoying your videos, music, or other content. These systems track huge amounts of data about everything people do online. They track which websites people visit. They track the topics they search. They track which devices people use to get online. This data lets them organize people into specific categories. Then companies can target those categories of people with their ads.

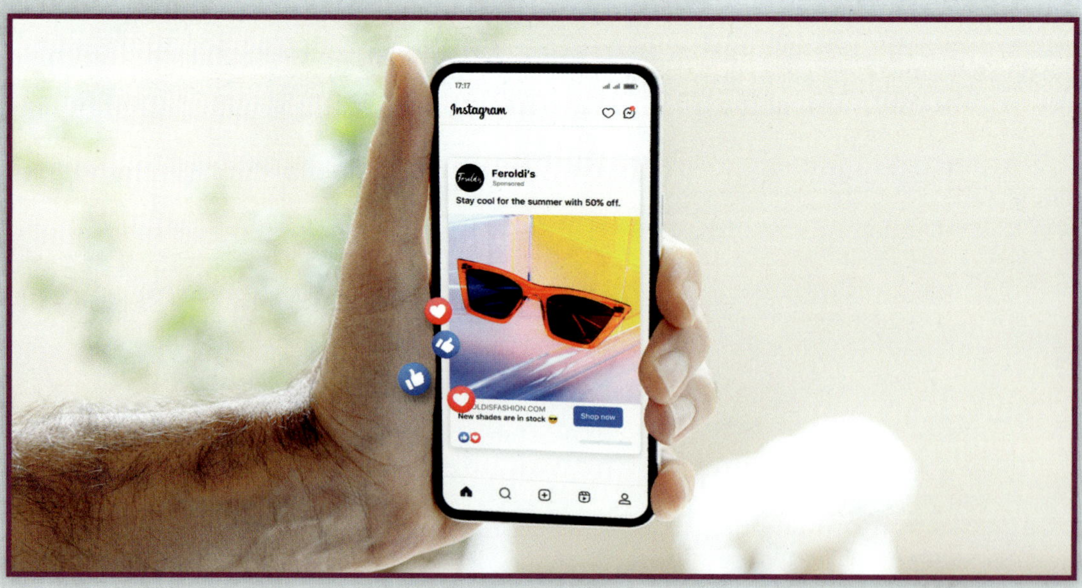

Targeted ads show consumers what they may be more interested in. The more clicks an ad gets, the more successful it is. Social media companies make money off of advertising.

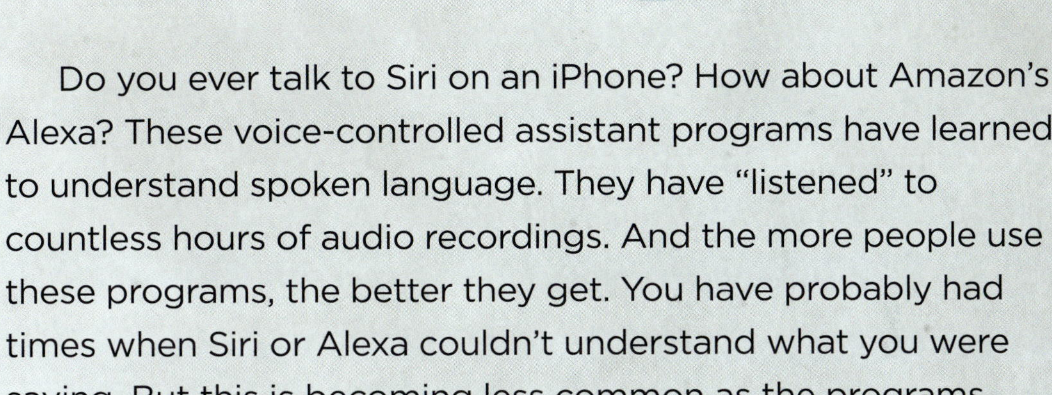

Do you ever talk to Siri on an iPhone? How about Amazon's Alexa? These voice-controlled assistant programs have learned to understand spoken language. They have "listened" to countless hours of audio recordings. And the more people use these programs, the better they get. You have probably had times when Siri or Alexa couldn't understand what you were saying. But this is becoming less common as the programs absorb more data and get smarter.

AI programs can also be trained to identify people based on their faces. First, **developers** show them huge numbers of photos of human faces. Then the AI starts to learn the things that set people's looks apart. This helps it recognize different faces, much like humans do. This technology can be used in all kinds of ways. Law enforcement uses it to try to identify people caught committing crimes on security footage. Modern smartphones can scan a user's face. This keeps others from using the device.

All of these AI applications are impressive. They are the result of years of research. Computer scientists and other experts work hard to build them. But they are still just the tip of the iceberg. All kinds of AI technology powers our world today.

AI AND GAMING

You have probably faced off against many computer-controlled video game enemies. These enemies are controlled by a type of AI. But most video game AI is different from the complex AI systems used in other products. Video game AI doesn't gather data. It doesn't get better over time from learning. Instead, it's based on a set of rules created by the game's developers.

The developers might program an enemy to duck behind cover when you fire a weapon at them. This makes video game AI somewhat predictable. Players can learn the AI's patterns. This lets them get better at beating it. But some developers are experimenting with more advanced AI systems in their games. One day, your opponents might be able to learn and adapt the same way you do. This could make future games more challenging and realistic.

Gamers may find video games harder to beat in the future with AI.

Chapter 2

HIGH-TECH HEALTH CARE

The healthcare industry has received a major boost from new AI technology in recent years. It helps improve how diseases are **diagnosed**. It also helps create new medications.

AI systems are skilled at analyzing data to find patterns and make predictions. This makes them good at figuring out if someone is likely to have a certain illness. AI systems can look at a patient's test results. They can then look at other health data on record. They may notice tiny details that a human doctor would have a hard time seeing. Many experts believe this will lead to illnesses being diagnosed earlier and more accurately. This should make the illnesses easier to treat.

People are increasingly using wearable devices to track their health data. These devices track heart rates, daily exercise, and other information. This data will make health AI systems even more powerful. One day,

AI can help find patterns in health data. These patterns could potentially help diagnose illnesses earlier.

your smart watch might be able to tell you when you are about to get sick. It might even be before you notice the symptoms.

AI could also make it much easier for patients to talk with their doctors. Some AI programs can understand speech and translate languages. This feature could let people describe symptoms to their doctors even if they don't speak the same language. And AI-powered virtual doctors could even answer basic health questions for patients. This would free up time for healthcare providers. They would have more time for patients with more serious conditions. These patients often require a lot of hands-on attention.

Researchers are also using AI to help develop better medications. AI programs are able to notice very specific patterns in large amounts of data. This means they can sometimes identify a specific version of an illness. For example, there are many forms of cancer. Different treatments are used to fight each kind. With AI, researchers hope to get even more specific. They want to identify new categories within the categories they already know. This will allow them to create medication that is most effective for each form of cancer.

Analyzing data with AI can offer doctors more information they need to diagnose specific types of cancer. AI could even help make more targeted medications.

Sometimes drug companies have a hard time making enough of a medication to meet the demand. Some patients have to wait to get their medication. They possibly might not get the medication at all. One reason for this can be that too much of a medication was shipped to one place. Not enough was shipped to another. With AI, companies hope to better predict when and where medications will be needed. This will let them make sure as many people as possible receive the treatment they need.

Experts predict that AI will continue to have a growing impact on health care as time goes on. This could help people live longer, healthier lives. It could help them avoid the worst outcomes of illnesses through early detection. And it could do this while decreasing the costs of health care.

AI could help pharmacies know when to order certain medications in order to have them on hand.

A ROBOT SURGEON?

If you needed surgery, would you let a robot perform the procedure instead of a human doctor? Today's robotic surgery systems are controlled by human doctors. They allow the doctors to perform surgeries using smaller or fewer cuts. Patients have less pain and bleeding. They heal faster.

These systems may be human-controlled today. But researchers are already looking for ways to improve them using AI. It might not be long until surgical robots are fully controlled by AI for some types of procedures.

Chapter 3

BIG BUSINESS

Many companies are finding innovative ways to use AI. These innovations help companies save money and become more **efficient**.

The manufacturing industry has long relied on automation to produce goods quickly and safely. Robotic workers can perform the same task over and over again. They never get tired, slow down, or make mistakes. But human workers are still needed to do certain jobs. These jobs are too complex for robots to perform. AI is starting to change that. It's helping robots learn to do tougher jobs. And robots are doing these jobs quickly. Manufacturers are also using AI to speed up their processes in other ways. For example, AI systems help predict when robots are going to break down. Workers can then repair them before there is an issue.

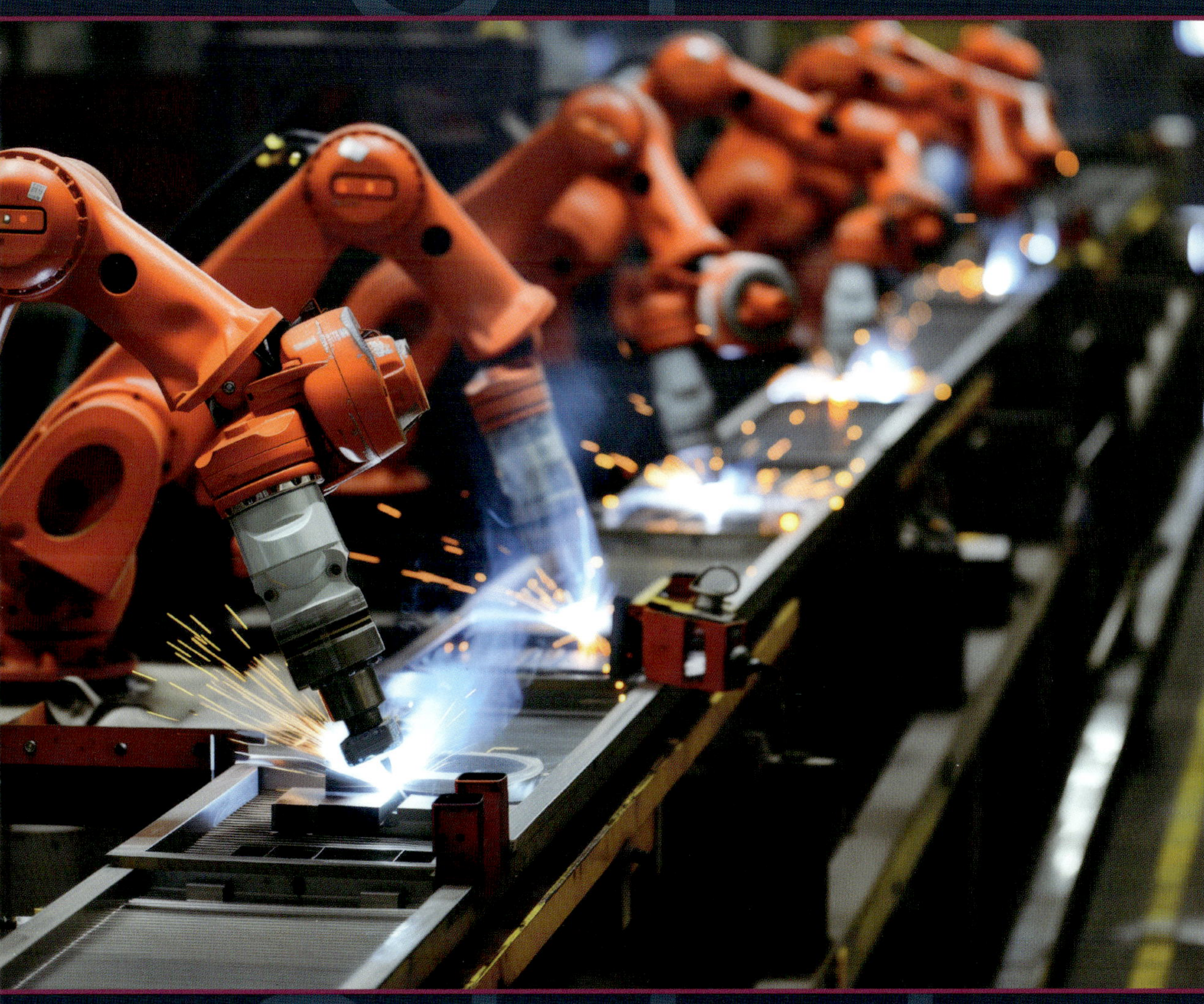

Robotic workers don't need breaks, and this makes them more efficient than human workers for some repetitive tasks. However, humans are still needed for more complex jobs, like managing the robotic workers.

AI also helps companies improve the way they distribute their products to stores. AI systems make it easier to keep track of **inventory**. This lets companies know exactly how much of a product to make. It helps them know how fast products are selling and where they are selling best. They don't make too much or too little. They can deliver products on time so stores always have stock to sell.

The vehicles that companies use to ship their products might soon be controlled by AI programs too. Self-driving vehicles combine AI with special cameras and sensors. This lets them "see" the road ahead, other vehicles, and other obstacles. They can then make decisions about how fast to go. They know when to turn and when to brake. Many companies hope to use these vehicles to transport goods.

Autonomous vehicles may be the delivery vehicles of the future.

AI also helps sell products. For example, AI-powered chatbots provide customer service. They can answer basic questions for customers about products. They can also help with complaints. They can organize returns and exchanges. Amazon and other companies have even created mostly automated, AI-driven stores.

In stores like these, shoppers simply take what they want. They put it in a cart and walk out. Cameras and sensors keep track of what customers take. The store automatically charges their accounts. There are still some human employees at these places, though. People are still needed to do other things in the stores.

WHAT ABOUT THE HUMANS?

As AI becomes more powerful and useful, many people are concerned that it will replace too many human workers. They fear that many of today's common jobs will eventually disappear.

Others argue that AI technology will create new kinds of jobs. For example, a human worker needs to oversee AI systems. AI could also simply change the way people work. People will need to learn new skills to use AI technology. But it's possible their jobs won't go away. Only time will tell what effect AI will have on the way people work. In the meantime, it is probably a good idea to stay informed about new AI technology and how it works. This knowledge could help you get a great job one day!

Chapter 4

GETTING CREATIVE

Generative AIs are some of the most talked-about AI programs in recent years. These programs create written and spoken language. They can make music, images, and other content. It can seem almost like magic. Want to see a realistic-looking picture of a purple, six-legged cat wearing sunglasses and riding a skateboard? Simply type the description into a generative AI program. It will probably give you a result pretty close to what you were imagining.

AI can also be a useful tool for human artists. For example, computer programs like Adobe Photoshop have AI-powered features. Imagine you took the perfect photo of a friend. But someone else ran by in the background just as you pressed the button. AI features could help you simply erase the unwanted person. AI would then make a guess at what the photo should look like if the person hadn't been there. It can then "fill in the blanks." The feature is easy to use. And the results are very realistic.

Many artists use AI as an aid when creating their own art.

These kinds of AI programs aren't limited to visual art. AI programs can also do interesting things with audio recordings. In 2023, legendary rock group the Beatles released their first new song in more than 25 years. Two of the group's four members had died years before. But all of them can be heard playing on the new song. AI technology helped the surviving band members and their recording team extract vocal and guitar parts from old recordings. These recordings were made when the other members were still alive. The old parts of the song were then combined with new parts recorded by the surviving members. The final song sounds like all four Beatles playing together again. AI helped give the band's fans something they had dreamed of for years.

Probably the most well-known AI systems in recent years are large language models (LLMs) such as ChatGPT. These programs can chat with users. They produce text that reads much like something a human would write. They can answer questions and hold conversations. They can also help users with their own writing.

A sound engineer using AI was able to help create the newest Beatles single.

LLMs are like many other AI systems. They are trained using huge amounts of data. Then they look for patterns. They don't "think" in the way humans do. Instead, they use the data they've absorbed to predict which word should follow the next. It is very similar to the way a phone will suggest words as you type a text. It's also like the way Google will guess what you want to search for based on the first few letters or words you type.

AI is already being used in so many ways. Can you imagine the things it will do in the future?

ASK AN AI

Let's go straight to the source and ask the AI text generator ChatGPT how AI works.

Hi, ChatGPT
Hello! How can I assist you today?

What are some of the most exciting uses of AI?
AI, or smart computer programs, can do lots of cool things! Imagine a robot friend that helps doctors find out if someone is sick really early, making it easier to help them feel better. AI can also make your toys and games even more fun by learning what you like and suggesting new things. It's like having a super-smart buddy that helps you learn new words and read stories. Cars can be like smart friends too, using AI to drive safely and avoid problems on the road.

Do you have a favorite kind of AI technology?
Well, I don't have feelings or favorites like people do, but I really enjoy helping with all sorts of things! I like being a friendly helper to answer questions, share interesting facts, and chat with you. So, my favorite "kind" of AI is being your helpful companion!

Thanks for sharing your thoughts!
You're welcome! If you have any more questions or if there's anything else I can help you with, feel free to ask. Have a great day!

(This conversation has been slightly edited and condensed for space.)

ACTIVITY:

A FRIENDLY CHAT

With an adult, try using AI to see its strengths and weaknesses.

1. Log on to ChatGPT and ask it to tell you a story about something. You can pick the topic.
2. Read what ChatGPT tells you. What do you think of it? What changes would you make to the story? Ask ChatGPT to make these changes.
3. Is the new version as good as you want it to be? Keep asking ChatGPT for more changes until you are happy with the results.
4. Now copy and paste the story into a program like Microsoft Word or Google Docs. Edit the story yourself to fix any details that ChatGPT didn't get right.

What did you think of using AI to help you write a story? Are you happy with the results? Was it easier than writing a story yourself? Or would it have been easier to avoid ChatGPT in the first place? What things did AI do well? What did it struggle with?

FIND OUT MORE

Books

Gitlin, Martin. *Smart Healthcare.* Ann Arbor, MI: Cherry Lake Publishing, 2021.

Gitlin, Martin. *Smart Vehicles.* Ann Arbor, MI: Cherry Lake Publishing, 2021.

Gregory, Josh. *Robots.* Ann Arbor, MI: Cherry Lake Publishing, 2018.

Schwartz, Heather E. *Medical Artificial Intelligence Breakthroughs.* Rochester, MN: Mayo Clinic Press, 2023.

On the Web

Search these online sources with an adult:

"Artificial Intelligence." Britannica for Kids.

"ChatGPT." OpenAI.

"Generative artificial intelligence facts for kids." Kiddle.

"Programming Robots." National Geographic.

"What is artificial intelligence (AI)?" IBM.

GLOSSARY

automate (AW-tuh-mayt) replace human workers with computer systems

data (DAY-tuh) information used to create, process, or support something

developers (dih-VEH-luh-purz) people who create AI and other computer programs

diagnosed (DYE-ig-nohsd) identified an illness

efficient (ih-FIH-shunt) able to do something using as few resources as possible, such as time

generative (JE-nuh-ruh-tiv) able to create something

inventory (IN-vuhn-tor-ee) record of the products a company has in stock

INDEX